Oil: A Natural Resource

Lisa Benjamin

Contents

Black Gold	2
How Was It Formed?	4
How Do We Find It?	8
How Do We Get It Out of the Ground?	12
How Do We Use It?	16
Glossary	20
Index	Inside back cover

Black Gold

Jet fuel, plastics, detergents, and spandex are all very different. But these products and materials have a common connection. Can you guess what it is? They're made from oil.

Products made from oil not only power airplanes, they also cover airport runways.

Oil, or **petroleum,** is a black liquid found at or beneath Earth's surface. It's thick, gooey, floats on water, and has a sharp smell. It's a valuable **natural resource** that has many important uses. In fact, oil has proven to be so valuable, it was nicknamed "black gold," after the yellow-colored metal.

The grass on this field is actually Astroturf, which is made from plastic that's produced from oil.

Products made from oil can be spun into synthetic fibers such as nylon and spandex. These fibers can then be used to make comfortable, flexible clothing.

Oil is used to make detergents like these, and the containers that hold them.

How Was It Formed?

Oil is a **fossil fuel.** So are coal and natural gas. All fossil fuels contain energy that comes from the sun. They began to form millions of years ago when lush plants and wet areas covered Earth. Coal formed on land, in swampy areas. Oil and natural gas formed at the bottoms of oceans and lakes.

After ancient plants and animals died, layers of **sedimentary rock** slowly formed on top of the remains. Sedimentary rock is soft. It was created from mud, sand, and silt. Over time, layers of rock built up, creating heat and pressure. This gradually turned the plant and animal remains into fossil fuels.

Some oil began to form during the Carboniferous period, about 350 million years ago.

Movements within Earth's crust caused rock layers to shift, slide, collide, and break. Most were pushed up. Ocean waters also moved. That is why some stores of petroleum can be found below areas of dry land today.

Oil and other fossil fuels are sources of energy. They are also **nonrenewable resources.** Unlike trees, which can be replanted, oil, coal, and natural gas can't be replaced, because they take several million years to form. So, once we've used them up, we won't be able to find more.

How Oil Was Formed

1. In ancient lakes and oceans, tiny plants and animals died and sank to the bottom.

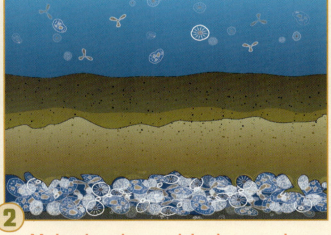

2. Mud and sand covered the decaying plant and animal remains.

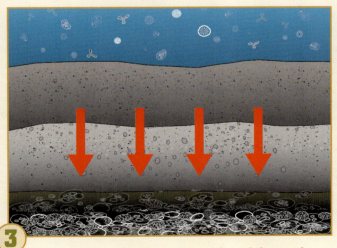

3. The mud and sand hardened and formed into layers of rock. Over time, more and more layers piled up, several miles deep, pressing down on the plant and animal remains.

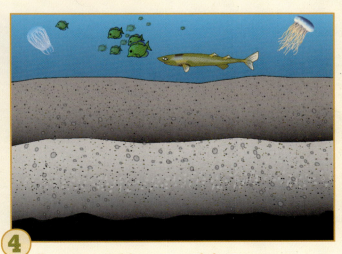

4. Pressure and heat turned the remains into oil. Beneath the rock layers, the temperatures were 212°F to 482°F.

Once the petroleum formed, most of it remained buried. Oil is fairly light in weight and some of it seeped up through tiny cracks in the rock layers. Pressure from sedimentary rock also pushed the liquid upward in places. However, layers of solid rock often blocked this flow and trapped the oil underground. The little oil that did rise to the surface collected in pools. Sunlight and weather changed the liquid into a thick substance that looks similar to dark syrup.

Quick Fact!

The word *petroleum* comes from the Latin words *petra*, which means "rock," and *oleum*, which means "oil." So, petroleum is "rock oil."

Petroleum in its natural state is **crude oil**, which contains water, other materials, and gases. It must be cleaned and treated before it can be used. Some crude oil is so thick it's almost solid. Some is found as a thin, runny liquid. Crude oil can also vary in color.

People have used oil for thousands of years. Most of it was found in pools that had seeped up to Earth's surface. Ancient people first used oil to light torches. Native Americans later used it as medicine and to waterproof canoes. In Europe, petroleum was made into ointments to treat back pain and bruises.

During the 1800s, it was discovered that a fuel called kerosene could be made from oil. Kerosene was burned in lamps to light homes and offices. Suddenly, the need for oil increased.

In ancient China, pipes made from bamboo stalks were used to suck oil from below ground. To break through the surface, a heavy chisel-like tool was repeatedly pounded into the ground.

At first, oil was mainly used to produce kerosene. Gasoline was considered a useless by-product. All of that changed when the automobile became popular during the early 1900s. Models, such as the one seen here, had engines that could be fueled with gasoline. The need for oil grew.

How Do We Find It?

As you know, most petroleum is trapped underground. Beneath the surface, there are folds within the sedimentary rock. These folds are similar to a throw rug bunched up in places on the floor, but on a much bigger scale. In these folds, oil has collected into huge pools, or deposits. But, it's impossible to see exactly where it's located.

Scientists called **geologists** try to find the oil. They examine basins of sedimentary rock where most deposits are found. The petroleum is buried far below the surface. The kinds of rocks in these basins are shale, sandstone, and limestone. They are about 3,000 to 15,000 meters thick (10,000 to 50,000 feet). They formed from sand and mud over millions of years.

This basin contains shale, the most common rock source for oil. When heat is applied, the rock can yield liquid petroleum. However, this process is expensive. Also, getting rid of the shale after the oil has been removed can harm the environment.

To find oil, geologists conduct surveys, or studies. First, they map areas where sedimentary rock is found. Using photographs from satellites and aircraft, they can make detailed observations about land areas and ocean floors. Next, they collect rock and soil samples from the surface to learn more about the layers below.

Geologists also set off small explosions to trigger sound waves that bounce off rock. They use equipment called seismographs to measure the sound waves, which help them learn about the structure of underground layers.

Once geologists have gathered all their information, they feed it into a computer to create a model. The model provides a picture of rock layers that helps geologists guess where oil deposits might lie.

Geologists not only search for new petroleum sources, they also look for ways to get more out of oil fields that have already been tapped. An oil field is an area where large deposits of petroleum are found.

Geologist Wayne Ahr is part of a project funded by the United States Department of Energy. He hopes to extract untapped oil that's been discovered, yet remains buried. He believes that a liquid detergent could be used to flush out the oil. This liquid would loosen oil from rock grains, just as dishwashing detergent removes grease from dirty dishes.

Like a sponge that soaks up water, sedimentary rock has pores, or tiny holes, that soak up oil. Can you find these holes in the top photo? Some of the oil sticks to the rock. No amount of pressure can push it out. To complete his study, Wayne Ahr first examines rock samples through a microscope. The images look like the magnification shown in the bottom photo.

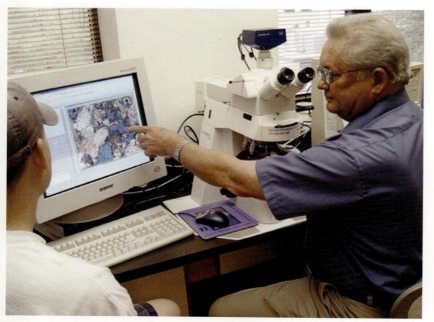

Wayne Ahr, shown here with a student, can load these magnifications onto a computer. This helps him track the size and shape of the rock pores. In the next part of the study, engineers will insert a gas such as carbon dioxide, or a liquid such as a detergent, into the ground to flush out the oil.

Investigate How Oil Deposits Form

You need a large, coarse sponge with lots of holes, a large bowl, a small, transparent plastic plate, and a ruler.

1 *Record* what you know about sedimentary rock. Is it soft or hard? Is it solid or does it have holes?

2 Use the ruler to *measure,* as you pour water into the bowl, up to the one-inch mark. The water represents oil that has formed underground.

3 Next, float the sponge on top of the water. It represents sedimentary rock, such as sandstone and limestone. How is it similar to these kinds of rock? What happens to the water now that the sponge is in the bowl?

4 Place the clear, plastic plate on top of the sponge. The plate should be a little bit smaller than the bowl. It represents a layer of solid rock. *Predict* what will happen to the water when you press down on the sponge, using the plate. Next, gently press down on the plate, putting pressure on the sponge.

5 *Observe* how the water changes. How has pressure from the plate and the sponge affected the water? How does pressure from layers of sedimentary and solid rock affect oil?

6 Where has some of the water collected? *Record* your observations. How is this similar to the way that oil deposits form? What can you *conclude* about oil fields and underground rock layers?

11

How Do We Get It Out of the Ground?

Once geologists identify a place where petroleum might be located, there's only one way to be sure if it's really there. They have to start drilling! They bring up **cores,** or rock samples, to the surface and test them to see if there's petroleum present. If there is, oil companies work to remove it from the ground.

To reach deposits, oil companies drill deep wells. Digging a well is a lot like inserting a straw into the ground. After a well has been dug, a pump is lowered to the bottom. It puts on pressure to push the oil up to the surface.

Above ground, towers called **derricks** provide supports for drills. A drill is made up of long lengths of pipe. On the end, there's a rotating cutting tool called a bit that can bore a hole through soft, sedimentary rock, but also through layers of hard rock.

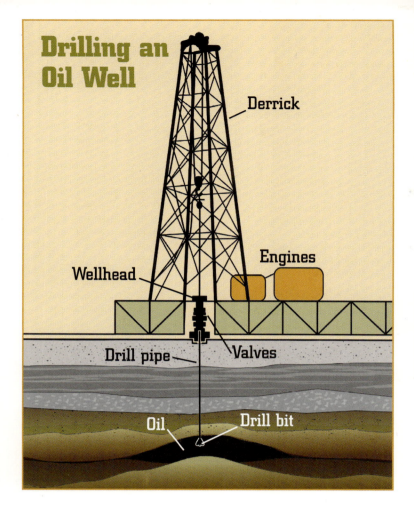

Drilling an Oil Well

Pressure from pumps causes oil to explode out of the ground in a great rush. Gushers like this one are very dangerous. So, valves are attached to wellheads to keep the oil in check.

Inventors at Work: Beginning an Oil Boom

In 1859, the first modern oil well was sunk by Edwin L. Drake in Titusville, Pennsylvania. Drake's methods influenced how petroleum is drilled today. Drake invented a long pipe, called a drive pipe, that could be battered into the ground. He built a wooden structure to support the pipe, just as steel derricks support modern-day drills. Drake used his drive pipe to drill almost 22 meters. This was actually a shallow well, but he struck oil nevertheless. Using a hand pump, he and his workers brought the oil to the surface. They collected it in a washtub! With the success of Edwin Drake's well, the petroleum industry was born.

About one-third of the world's petroleum comes from oil fields buried under the ocean floor. Most are located in the Persian Gulf, the Gulf of Mexico, and the North Sea between Scotland and Norway. To reach them, petroleum companies set up **offshore oil rigs.** These structures are as tall as skyscrapers. They stand on large platforms on the water.

Two kinds of rigs—jack-ups and semisubmersibles—are used to search for oil fields. They rest on floating platforms that can be moved to different sites as the search continues. Once the oil has been discovered, more permanent, or fixed, platforms are set up to pump the oil from the bottom of the sea.

Offshore workers use the same techniques to drill a well on the ocean floor as oil workers do on land. Jack-up rigs, such as the one shown above on the left, search for petroleum in water 75 meters deep or less. Semisubmersibles, such as the one above on the right, are used in greater depths. The platform rests on tanks partially filled with water. They help keep the rig steady against high winds and waves in deep waters.

Drilling for oil can be dangerous work. It can also be harmful to the environment. Leaks from wells drilled close to the shore can create oil slicks. This pollutes beaches and endangers sea life. Also, after a land area has been drilled, it must be restored before it can be used for other purposes, such as farming.

This rescue worker is tending to a bird with feathers coated in petroleum from an oil spill.

Oil Around the World

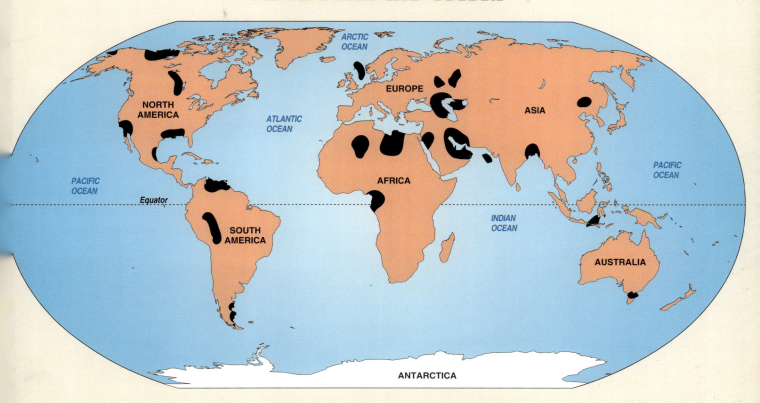

The world's main oil-producing areas are the Middle East, North America, Western Europe, and Northern Africa.

How Do We Use It?

Once petroleum has been pumped from the ground, it's shipped from oil fields to countries all over the world. Oil companies use ships called tankers and pipeline systems to send crude oil to far-off destinations. During shipping, they must take care to avoid accidental oil spills, which damage the environment.

The Trans-Alaska Pipeline connects Prudhoe Bay to Valdez and runs about 800 miles long.

Quick Fact!

There are three major kinds of crude oil and within each kind are different grades, or types. Up to 100 different grades of crude oil are drilled and traded around the world.

At **refineries,** or factories, the crude oil is used to make a number of products. First, it's cleaned with chemicals. Then it's pumped into a furnace, where heat turns the liquid petroleum into a vapor. This vapor is then piped into a tall piece of equipment called a distillation tower where it's separated into different materials.

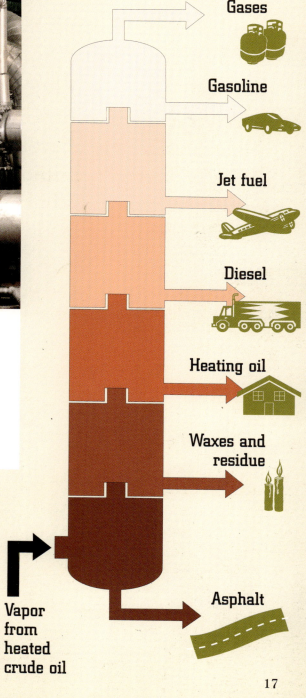

Turning Up the Heat on Crude Oil

An oil refinery can contain miles of pipes that are used to pump crude oil to different tanks and towers where it's cleaned and treated.

Inside a distillation tower, the heated vapor begins to rise. The lightest oil vapors—for instance, gasoline—rise to the top, while heavier ones—such as heating oil—stay near the bottom. Each different kind of vapor cools and condenses back into liquid form. The different liquids collect in trays along the tower.

17

In the United States we use almost 20 million barrels of oil a day. One barrel contains 42 gallons. Burning oil products releases carbon dioxide. This gas remains in the atmosphere, where it traps heat and raises the temperature of Earth's surface. This is known as global warming. To help cut back on the use of oil, scientists now search for alternative sources of fuel.

Scientists also look for ways to use oil more carefully. If we conserve oil now, it will be available for future generations. This is important because petroleum has greatly affected how we travel, what we wear, and how we work and play. In fact, it's hard to imagine our world without it. What can you spot around you that's made from oil?

Trucks deliver heating oil and other fuels from refineries to communities across the country.

In addition to separating crude oil into different parts, refineries also produce a variety of **petrochemicals.** These chemicals are used to make a wide array of materials and products. Detergents, medicines, insecticides, and weed killers are all produced from petrochemicals. Pens, crayons, and paints are too.

Plastic is also made from petrochemicals. It can be shaped into thousands of products, from flexible clothing to these sturdy rides at the Legoland theme park in California. It's important to reuse plastic products because they don't decompose, or break down. If they're thrown away, they pile up in landfill areas.

Young Environmentalists at Work: Don't Be Crude

Motor oil is another product made from crude oil. It helps engines run smoothly. Once it's been used, though, it must be disposed of in the right way. Three young girls from Texas learned that old motor oil was being used to kill weeds. When this happens, the oil seeps into the ground, where it can contaminate the water supply.

So the three girls, Barbara Brown, Lacy Jones, and Kate Klinkerman, who are shown on the right, started an organization called *Don't Be Crude*. Their goal was simple. They wanted to educate their community about the proper way to get rid of motor oil. They set up bins at gas stations and local stores to collect bottles of used oil. The motor oil was then recycled and used for asphalt on Texas roads. For their efforts, the girls won the Environmental Youth Award, which was presented at the White House.

Glossary

core (KOR) a sample of rock that geologists remove from underground to test for the presence of oil

crude oil (KROOD OYL) petroleum in its natural state

derrick (DER-ik) supporting structure of an oil well

fossil fuel (FAH-sul FYOOL) a source of energy that formed millions of years ago from dead plants and animals

geologist (jee-AH-luh-jist) a scientist who studies the rocks on and under Earth's surface

natural resource (NA-chuh-rul REE-sors) a material found in nature that people can use to suit their needs

nonrenewable resource (nahn-rih-NOO-uh-bul REE-sors) an energy source that, once used, cannot be replaced

offshore oil rig (AWF-shor OYL RIG) a platform used to drill oil wells on the ocean floor

petrochemical (peh-troh-KEH-mih-kul) a chemical made from refined crude oil and used to make products such as plastics, medicines, and fertilizer

petroleum (puh-TROH-lee-um) a fossil fuel, also known as oil

refinery (rih-FYE-nuh-ree) a factory where crude oil is processed with heat and chemicals in order to be separated into usable parts

sedimentary rock (seh-duh-MEN-tuh-ree RAHK) soft, porous rock that formed from mud, sand, or silt